地 球

我们在太阳系里的家

EARTH

Our Home in the Solar System

（英国）埃伦·劳伦斯／著　刘 颖／译

江苏凤凰美术出版社

著作权合同登记图字：10-2022-144

图书在版编目（CIP）数据

地球 : 我们在太阳系里的家 / (英) 埃伦·劳伦斯著 ; 刘颖译. -- 南京 : 江苏凤凰美术出版社, 2025. 1. -- (环游太空). -- ISBN 978-7-5741-2027-3

Ⅰ. P183-49

中国国家版本馆CIP数据核字第2024MR8872号

策　　划　朱　婧
责任编辑　高　静　奚　鑫
责任校对　王　璇
责任设计编辑　樊旭颖
责任监印　生　嫄
英文朗读　C.A.Scully
项目协助　邵楚楚　乔一文雯

丛 书 名　环游太空
书　　名　地球：我们在太阳系里的家
著　　者　（英国）埃伦·劳伦斯
译　　者　刘　颖
出版发行　江苏凤凰美术出版社（南京市湖南路1号　邮编：210009）
印　　刷　南京新世纪联盟印务有限公司
开　　本　710 mm × 1000 mm　1/16
总 印 张　18
版　　次　2025年1月第1版
印　　次　2025年1月第1次印刷
标准书号　ISBN 978-7-5741-2027-3
总 定 价　198.00元（全12册）

营销部电话：025-68155675　营销部地址：南京市湖南路1号
江苏凤凰美术出版社图书凡印装错误可向承印厂调换

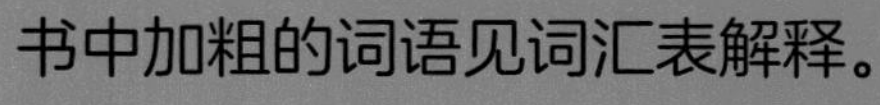

书中加粗的词语见词汇表解释。

Words shown in **bold** in the text are explained in the glossary.

太空里的神奇家园

Our Amazing Home in Space

想象一下，你正坐在一艘太空飞船上，等待从地球发射。

Imagine you are aboard a spacecraft waiting to blast off from Earth.

你听到倒计时：“3、2、1、0——起飞！”

You hear the words, “three, two, one, zero, and lift-off!”

当你的太空飞船被一枚巨大的火箭送向天空时，它摇晃得很厉害。

Your spacecraft shakes as it is carried into the sky by a huge rocket.

很快，你以每小时几万千米的速度在太空中飞行。

Soon you are speeding through space at tens of thousands of kilometers per hour.

你向窗外望去，看到一个由浅蓝色、绿色和棕色组成的世界。

You look out of the window and see a bright blue, green, and brown world.

这是地球，我们在太空里的神奇家园。

It’s **planet** Earth, our amazing home in space!

这张地球照片是2019年由人造卫星拍摄的。想想看，当照片被拍下时，你正在这颗美丽星球的某个地方！

This photo of Earth was taken by a **satellite** in 2019. Just think, you were somewhere on this beautiful planet when the photo was taken!

太阳系 The Solar System

地球以超过107 000千米每小时的速度在太空移动。

Earth is moving through space at over 107,000 kilometers per hour.

它围绕着太阳做一个巨大的圆周运动。

It is moving in a huge circle around the Sun.

地球是围绕太阳公转的八大行星之一。

Earth is one of eight planets circling the Sun.

八大行星分别是水星、金星、地球、火星、木星、土星、天王星和海王星。

The planets are called Mercury, Venus, Earth, Mars, Jupiter, Saturn, Uranus, and Neptune.

冰冻的彗星和被称为“小行星”的大型岩石也围绕着太阳公转。

Icy **comets** and large rocks, called **asteroids**, are also moving around the Sun.

太阳、行星和其他天体共同组成了“太阳系”。

Together, the Sun, the planets, and other space objects are called the **solar system**.

太阳系中的大多数小行星都集中在被称为“小行星带”的环状带中。

Most of the asteroids in the solar system are in a ring called the asteroid belt.

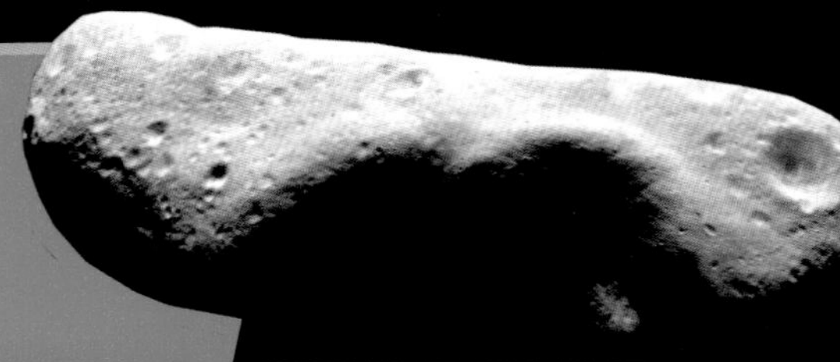

小行星 **An asteroid**

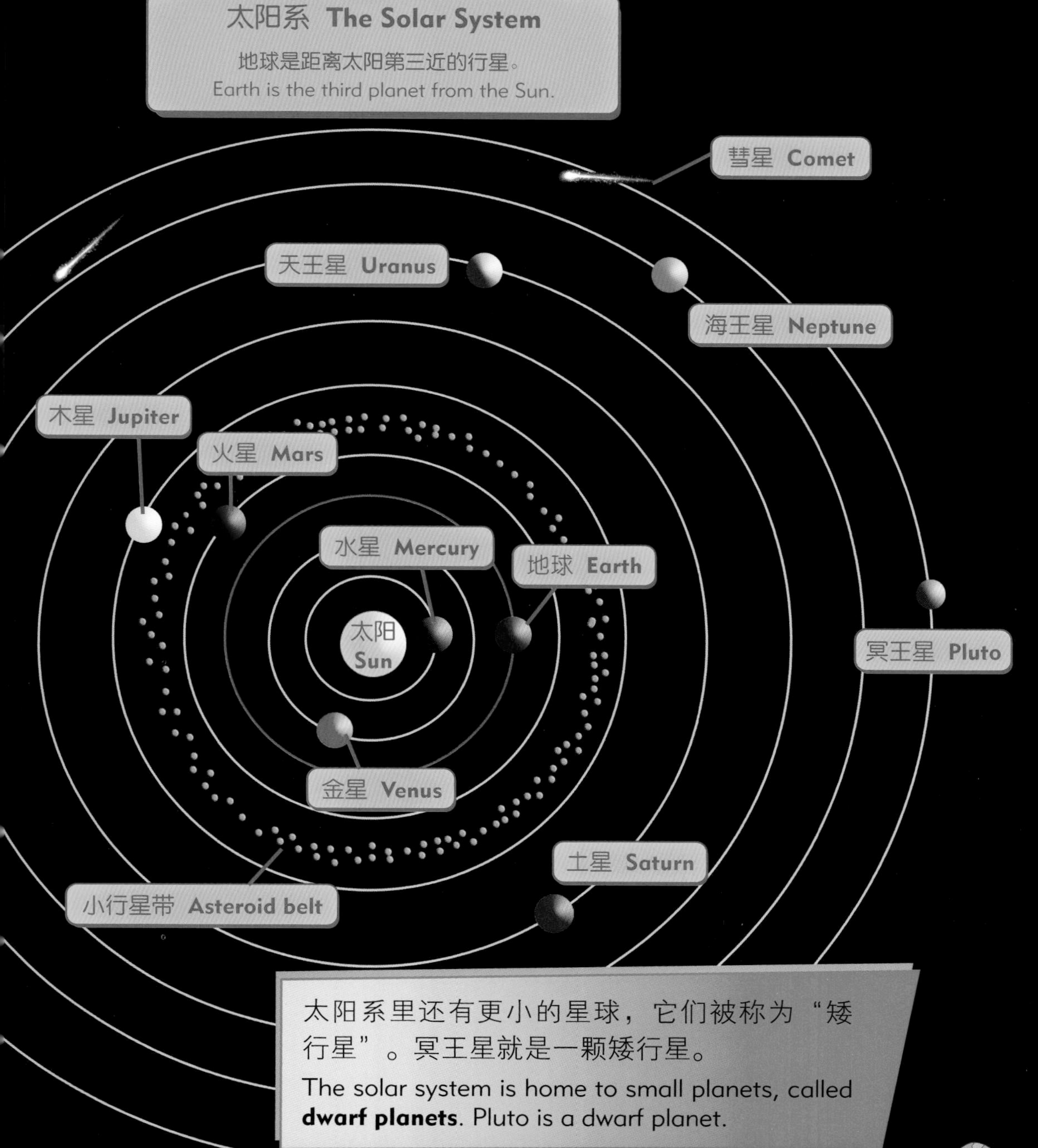

太阳系里还有更小的星球，它们被称为“矮行星”。冥王星就是一颗矮行星。

The solar system is home to small planets, called **dwarf planets**. Pluto is a dwarf planet.

地球的奇幻之旅
Earth's Amazing Journey

行星围绕太阳公转一圈所需的时间被称为“一年”。

The time it takes a planet to **orbit**, or circle, the Sun once is called its year.

地球绕太阳公转一圈需要略多于365天。

It takes Earth just over 365 days to orbit the Sun.

所以地球上的一年有365天。

So a year on Earth lasts 365 days.

绕太阳公转一圈是段很长的路程。

To orbit the Sun once, Earth must travel a long distance.

地球得走过9.4亿千米。

It makes a journey of about 940 million kilometers.

水星、金星、地球和火星是距离太阳最近的4颗行星。这张图显示了它们的大小对比。

Mercury, Venus, Earth, and Mars are the four planets closest to the Sun. This picture shows their sizes compared to each other.

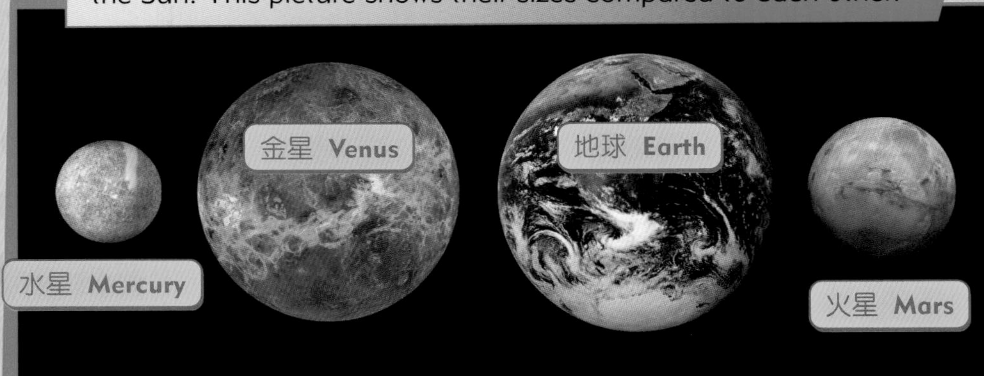

地球环绕太阳的旅程
Earth's Journey Around The Sun

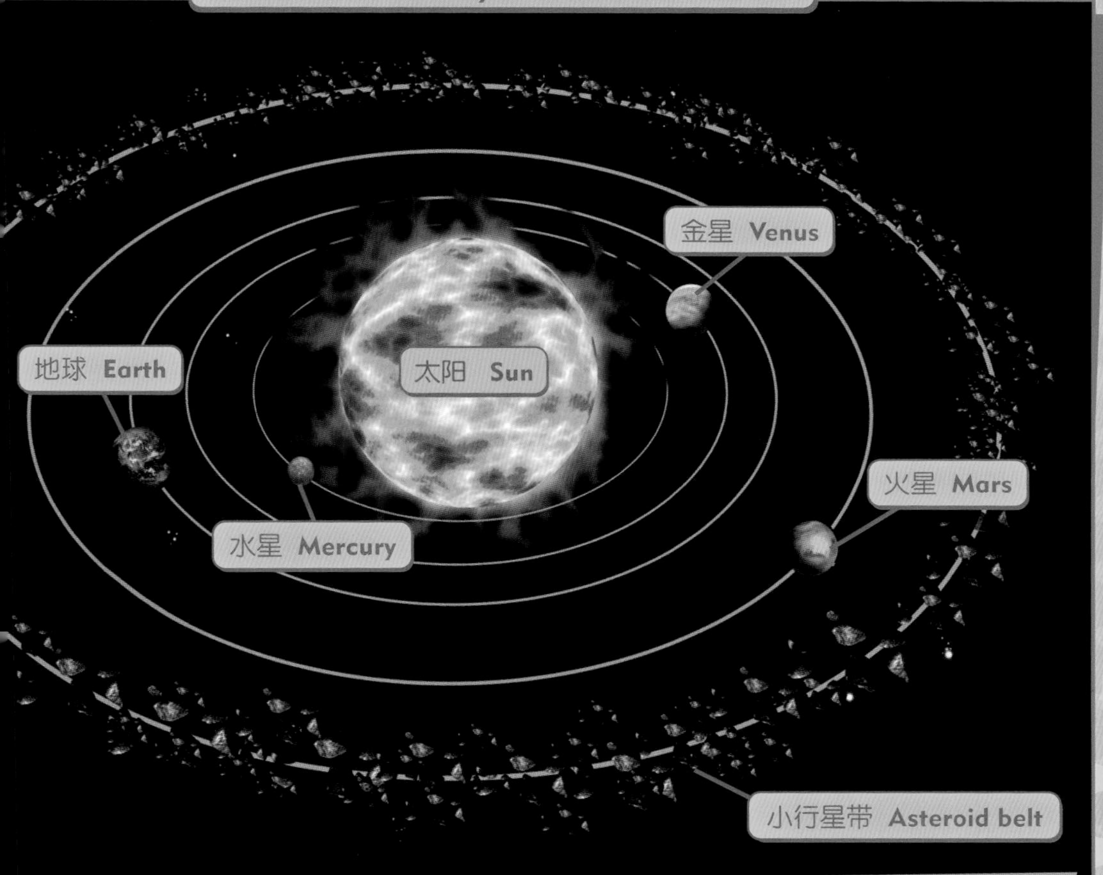

你可能认为，在读这本书时，你是一动不动的。但事实上，你和地球上的一切都在围绕太阳高速旋转！

You might think you are sitting still reading this book. In fact, you and everything else on Earth are zooming at high speed around the Sun!

在太空旋转

Spinning Through Space

当行星围绕太阳公转时，它也像陀螺一样自转着。

As a planet orbits the Sun, it also spins, or **rotates**, like a top.

地球自转一圈需要24小时。

It takes Earth 24 hours to rotate once.

地球自转是昼夜更替的原因。

The reason we have day and night is because Earth is spinning.

当你生活的地方面向太阳时，就是白昼。

When the place where you live faces toward the Sun, it is daytime for you.

当地球继续旋转，你生活的地方远离太阳时，黑暗降临，此时已是夜晚。

As it spins away from the Sun's light, darkness falls and it is night.

地球自转时，略向一侧倾斜。

Earth is slightly tilted to one side as it spins.

这张图显示了地球上的白昼和黑夜。地球朝向太阳的一面是白昼，而另一面是黑夜。

This picture shows how day and night look on Earth. It's daytime on the half of the planet that's facing the Sun. On the other half, it's nighttime.

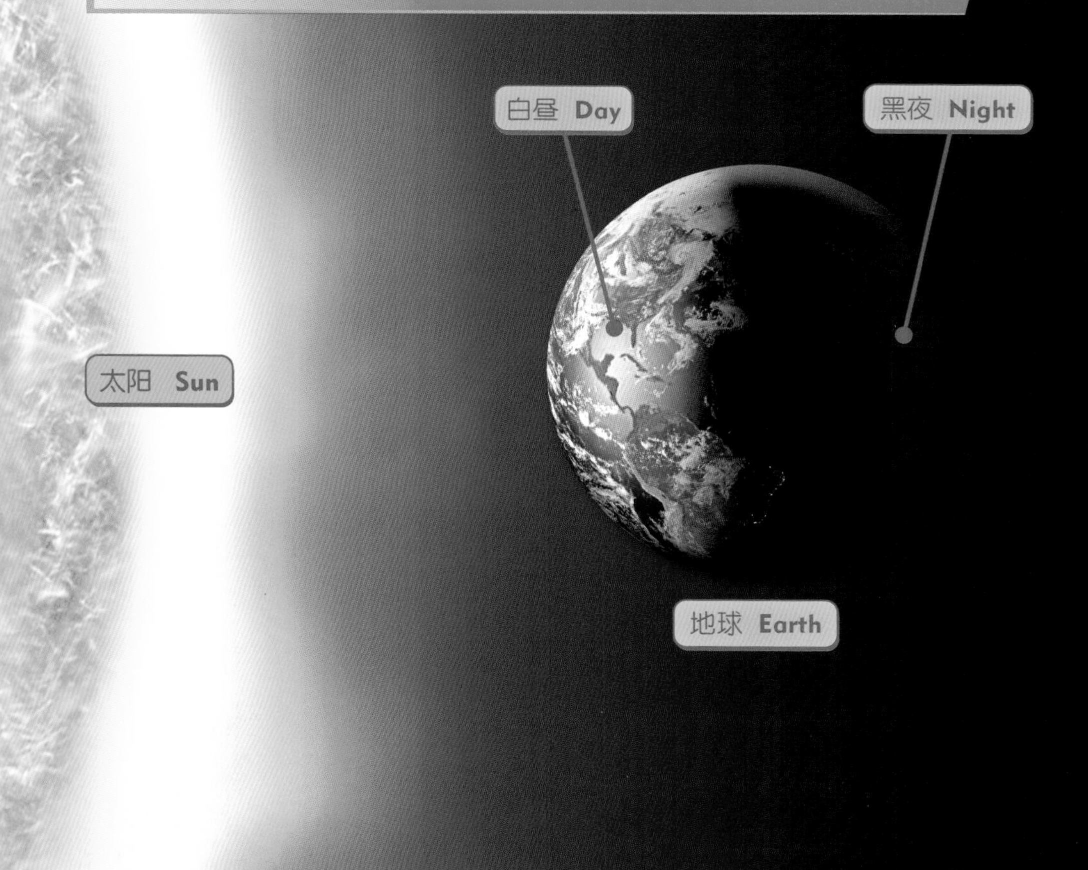

在这张图里，地球和太阳靠得很近。事实上，地球距离太阳约1.5亿千米。

This picture shows Earth and the Sun very close together. In real life, Earth is about 150 million kilometers from the Sun.

地球的“近邻”
Earth's Space Neighbor

地球和它的“近邻”月球一起在太空中穿行。

Planet Earth travels through space with its closest neighbor, the Moon.

月球围绕地球公转。

The Moon is orbiting Earth.

月球绕地球公转一圈需要略多于27天。

It takes the Moon just over 27 days to orbit Earth once.

当我们看见天空中的月亮时，它通常是白色或灰白色的。

When we see the Moon in the sky, it looks white or grayish-white.

这是因为太阳光照在月球上，又从月球上反射到了人眼中。

That's because the Sun's light is shining on the Moon and lighting it up.

这张图显示了月球与地球的大小对比。

This picture shows the Moon's size compared to Earth's.

月亮的形状看上去好像总在变化，但实际上并没有。当月球绕地球公转时，我们可以看到的是它被照亮的部分而已。

The Moon looks as if it changes shape, but it doesn't really. As the Moon travels around Earth, we see different parts of its shining surface.

这张图显示了我们在地球上看到的形状各异的月亮。

These photos show some of the different ways that we see the Moon from Earth.

这是月球表面的特写照片。

This is a close-up photo of the Moon's surface.

近距离观察地球
A Closer Look at Earth

如果将地球切成两半，你会发现地球分为好几层。

If you could cut Earth in half, you'd see that it is made up of different layers.

地球的外层是一层薄薄的岩石外壳，称为“地壳”。

The outer layer of Earth is a thin, rocky crust.

地壳下面是一层厚厚的岩石，由于温度极高，像柔软的太妃糖一样。

Next is a thick layer of rock that is so hot it has become soft, like toffee.

地球的中心是一个实心金属球，被一层高温液态金属包裹。

At the center of the Earth is a solid metal ball, surrounded by a super-hot layer of liquid metal.

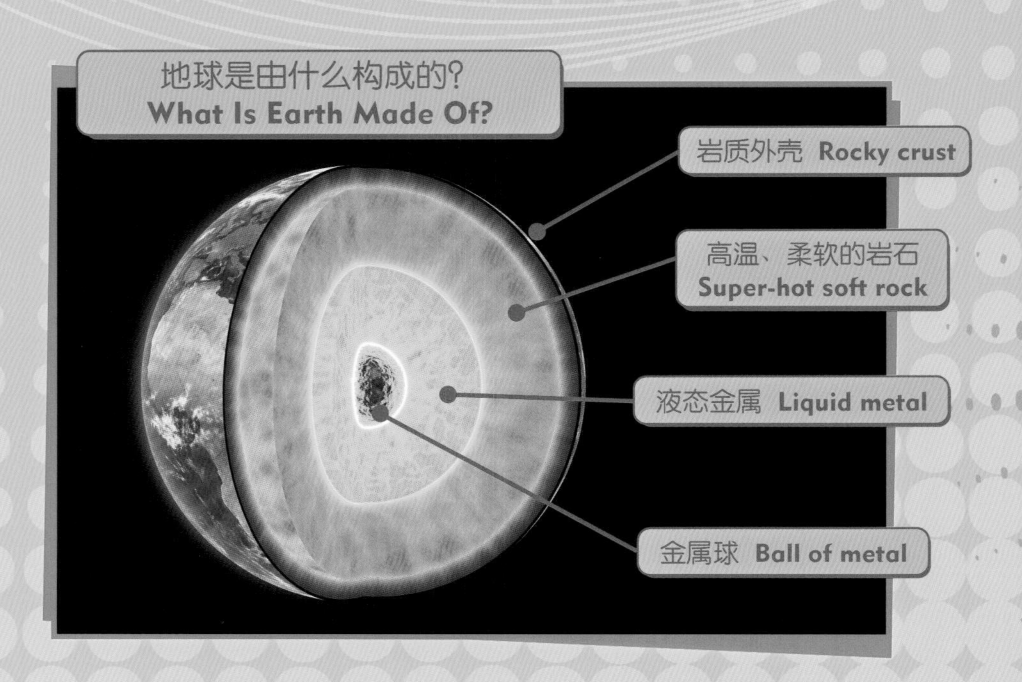

地球的周围有一层厚厚的气体，叫作“大气层”。
照片里是从太空看到的地球大气层。

All around Earth, there is a thick layer of **gases** called an **atmosphere**. This picture shows how the atmosphere looks from space.

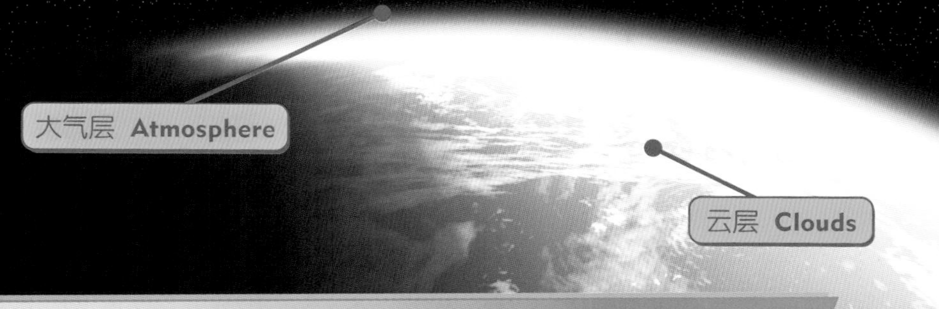

大气中含有氧气，这是人类和其他动物呼吸所必需的气体。

The atmosphere contains **oxygen**, which is the gas that humans and other animals need to breathe.

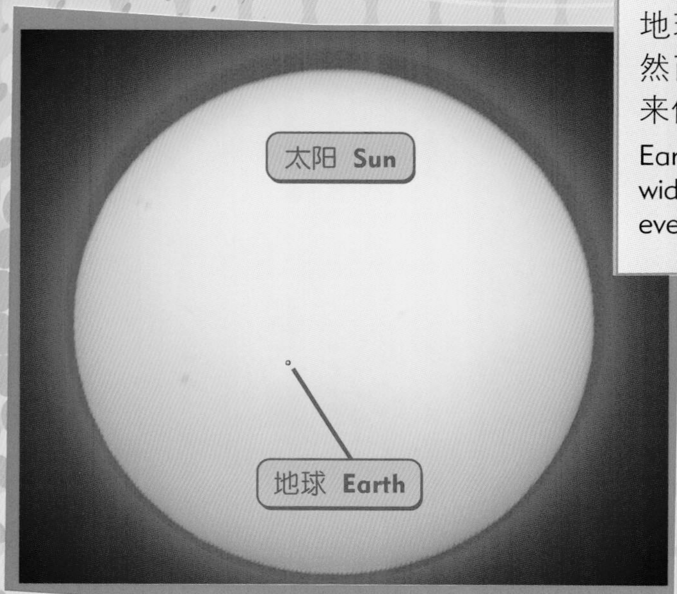

地球的直径约12 800千米。然而，与太阳相比，它看起来像一个小点！

Earth is nearly 12,800 kilometers wide. It looks like a tiny dot, however, compared to the Sun!

蓝色星球

The Blue Planet

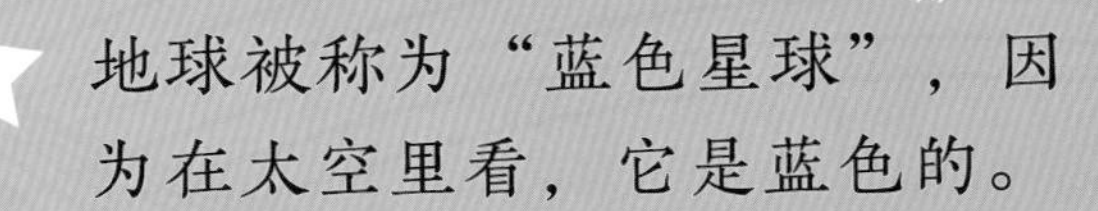

地球被称为“蓝色星球”，因为在太空里看，它是蓝色的。

Earth has been nicknamed “The Blue Planet”, because it looks blue from space.

这是因为地球表面约$\frac{3}{4}$的地方被水覆盖。

That’s because about three-quarters of its surface is covered by water.

海洋、湖泊和河流里有液态水。

There is liquid water in oceans, lakes, and rivers.

在北极和南极等极寒之地也有冰冻的水。

There is also frozen water in icy, super-cold places such as the North and South Poles.

科学家在太阳系的其他行星和卫星上也发现了冰和水。

Scientists have discovered ice and water on other planets and moons in the solar system.

然而，在已知行星和卫星中只有地球表面始终有液态水。

Earth is the only planet or moon that has liquid water on its surface all the time, though.

湖泊 A lake

地球上的海洋有数千千米宽。在某些地方，海洋的深度超过6.4千米。

Earth's oceans are thousands of kilometers wide. In places, they are over 6.4 kilometers deep.

这张照片显示了南极被冰雪覆盖的地面。有些地方的冰层厚达4.8千米。

This photo shows the ice-covered ground at the South Pole. In places, the ice is 4.8 kilometers thick.

特别的星球

A Very Special Planet

在很大程度上，地球是一颗非常特别的行星。

Earth is a very special planet in one important way.

它是太阳系中唯一已知有生命存在的行星。

It is the only planet in the solar system where we know there is life.

科学家认为地球上有超过800万种不同种类的生物。

Scientists think there are over 8 million different types of living things on Earth.

地球是植物、动物和微生物的家园。

Our planet is home to plants, animals, and tiny living things such as **microbes**.

它也是一种高等智慧生物的家园。说的就是你！

It is also home to one super-intelligent type of animal. That's you!

微生物 **Microbe**

这是一张微生物的特写照片。微生物太小了，只有用显微镜才能看到它们。

This is a close-up photo of a microbe. Microbes are so small, we can only see them with a microscope.

地球上有数十万种不同种类的动物和植物。

There are hundreds of thousands of different types of plants and animals on Earth.

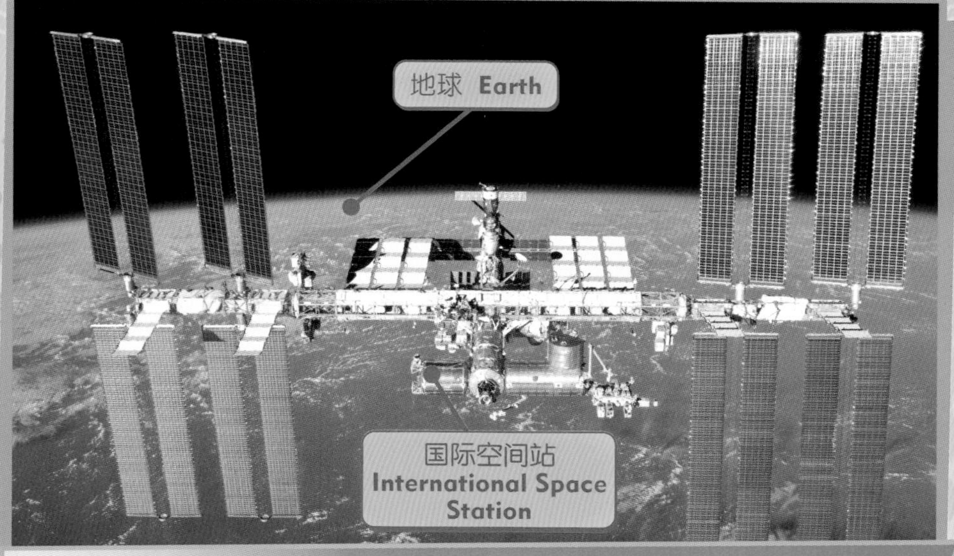

人类非常聪明，他们发明并建造了环绕地球公转的国际空间站。宇航员在那里生活和工作。

Humans are so smart they invented and built the International Space Station that orbits Earth. **Astronauts** live and work there.

有趣的地球知识
Earth Fact File

以下是一些有趣的地球知识：地球是距离太阳第三近的行星。

Here are some key facts about Earth, the third planet from the Sun.

地球是如何得名的
How Earth got its name

地球的英文单词“Earth”起源已久，原本的意思是“地面”（ground）。

The word "Earth" is a very old word for "ground".

行星的大小
Planet sizes

这张图显示了太阳系八大行星的大小对比。

This picture shows the sizes of the solar system's planets compared to each other.

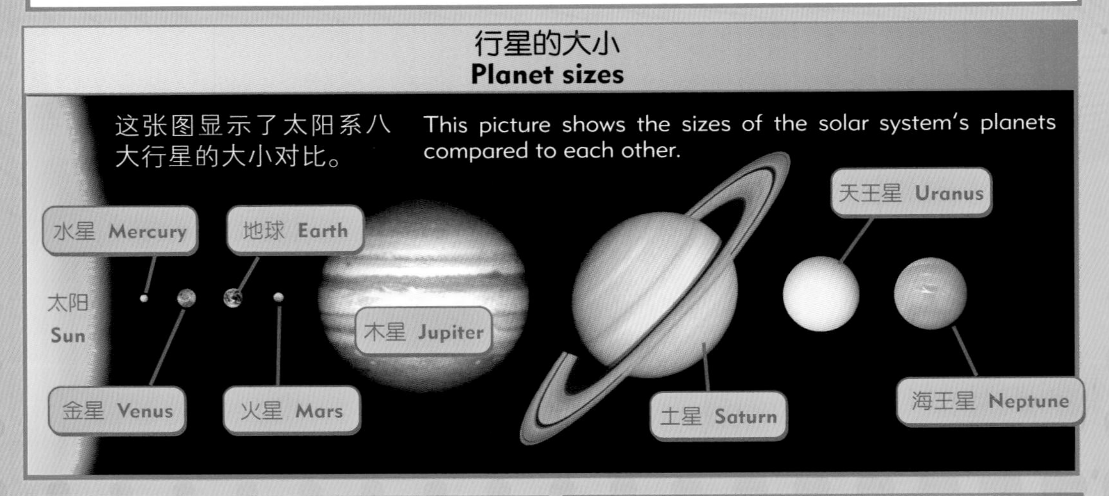

地球的大小
Earth's size

地球的直径约12 742千米

About 12,742 km across

地球自转一圈需要多长时间
How long it takes for Earth to rotate once

将近24小时

Nearly 24 hours

地球与太阳的距离
Earth's distance from the Sun

地球与太阳的最短距离是147 098 291千米。

地球与太阳的最远距离是152 098 233千米。

The closest Earth gets to the Sun is 147,098,291 km.

The farthest Earth gets from the Sun is 152,098,233 km.

地球绕太阳轨道的长度
Length of Earth's orbit around the Sun

939887974千米

939,887,974 km

地球围绕太阳公转的平均速度
Average speed at which Earth orbits the Sun

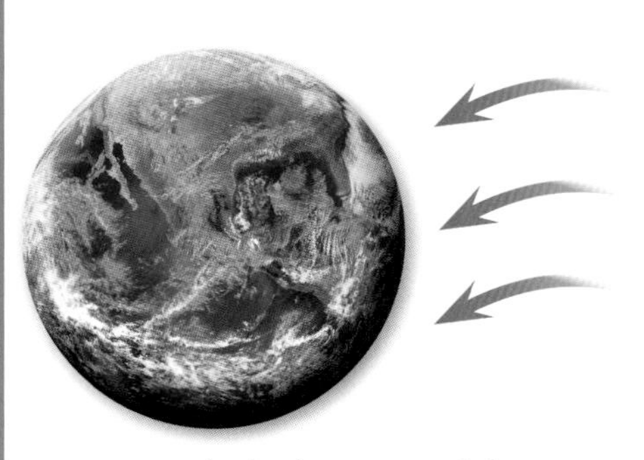

每小时10 7218千米

107,218 km/h

地球上的一年
Length of a year on Earth

略多于365天

Just over 365 days

地球的卫星
Earth's Moons

地球有1颗卫星。

Earth has one moon.

地球上的温度
Temperature on Earth

最高温度：58摄氏度
最低温度：零下88摄氏度
Highest: 58°C
Lowest: -88°C

动动手吧：太阳系成员大集合游戏

Get Crafty : Solar System Concentration Game

1. 将一大张纸裁成30张矩形卡片。
2. 选出15个天体，将每个天体画在一对卡片上。你可以选择太阳、月亮、八大行星、小行星和彗星等。
3. 将天体的名字也写在卡片上。
4. 打乱卡片的顺序，面朝下摆放，游戏开始啦！小朋友们互相抽取卡片，翻开后读出卡片上天体的名称。

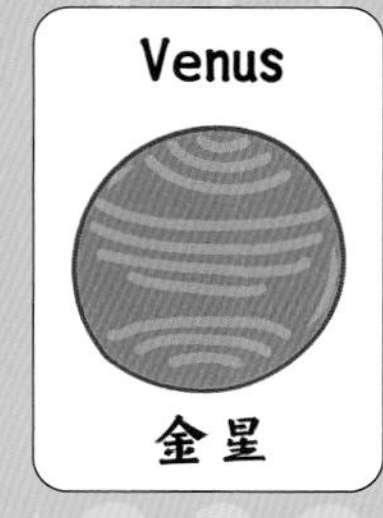

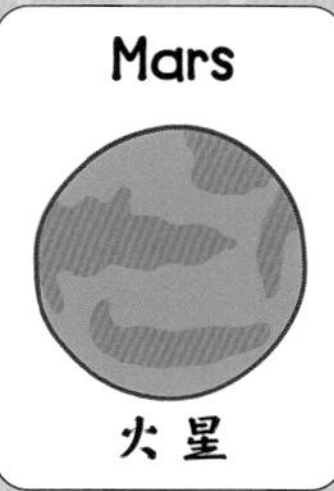

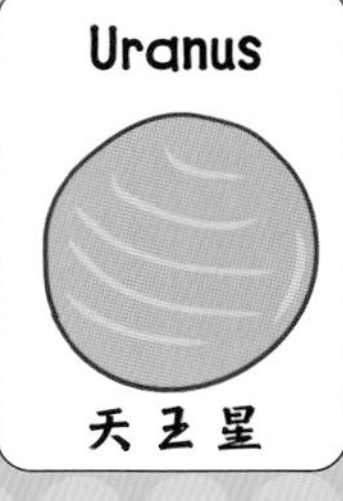

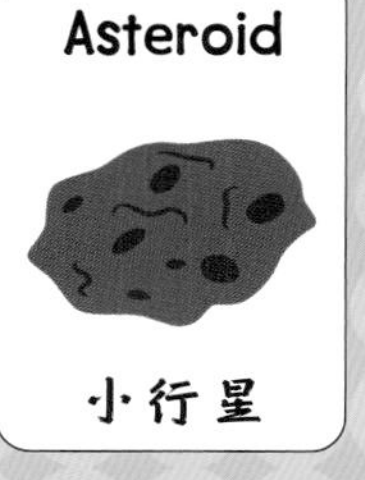

词汇表 Glossary

小行星 | asteroid

围绕太阳公转的大块岩石，有些小得像辆汽车，有些大得像座山。

宇航员 | astronaut

受过特殊训练，乘坐宇宙飞船进入太空的人。

大气层 | atmosphere

行星、卫星或恒星周围的一层气体。

彗星 | comet

由冰、岩石和尘埃组成的天体，围绕太阳公转。

矮行星 | dwarf planet

围绕太阳运行的圆形或近圆形天体，比八大行星小得多。

气体 | gas

无固定形状或大小的物质，如氧气或氦气。

微生物 | microbe

极小的生物，无法用裸眼看见。能够使人生病的细菌就是一种微生物。

公转 | orbit

围绕另一个天体运行。

氧气 | oxygen

空气中一种无形的气体，是人类和其他动物呼吸所必需的。

行星 | planet

围绕太阳公转的大型天体：一些行星，如地球，主要是由岩石组成的；其他的行星，如木星，主要是由气体和液体组成的。

自转 | rotate

物体自行旋转的运动。

人造卫星 | satellite

围绕行星公转的人造天体。人造卫星能向全球收发电视和手机信号。

太阳系 | solar system

太阳和围绕太阳公转的所有天体，如行星及其卫星、小行星和彗星。